"I have long admired Bill Crain's work with farmed animals, and I have drawn upon his extensive experiences with these sentient beings in my own research and writing. In his latest book, *Animal Stories: Lives at a Farm Sanctuary*, Bill clearly shows that if we approach nonhuman animals with an open mind, as well as with a warm and welcoming heart, they will surprise us with their compassion, empathy, rich and deep emotional lives, and intelligence. I strongly recommend this delightful book to all who want to learn more about them, and perhaps themselves."—**Marc Bekoff**, Professor Emeritus of Ecology and Evolutionary Biology, the University of Colorado and author, *The Emotional Lives of Animals, The Ten Trusts* (with Jane Goodall), and *Rewilding Our Hearts*

"William Crain shares the deeply moving stories of rescued farm animals, who before rescue, were likely destined for slaughter. The narratives showcase each animal's individuality, personality, courage, resilience, compassion, and playfulness, as well as their deep attachments with other animals and caring humans. The stories remind us of the sentience, intelligence, and mystery of non-human animals, touching something deep inside of our humanity. Indeed, our connections with animals and the natural world impact our own emotional and physical well-being in a multitude of ways. As you read *Animal Stories*, you will experience moments of laughter and tears, as well as an appreciation for those who, like the author and his wife, create sanctuary."—**Angela Crawford, Ph.D.**, Licensed Psychologist, co-author, The Behavioral Medicine Treatment Planner

"The tales in *Animal Stories* are delightful. In the book, you'll meet Sprinkles, a sheep who wasn't sheepish, Charlotte, a lame chicken, who became friends with a goat named Violet, and many more. Crain's love and care for these rescued animals infuse every page."—**Charlotte L. Doyle**, Professor of Psychology, Sarah Lawrence College and author, *The Creative Process: Stories from the Arts and Sciences*

"Bill Crain invites us into a world that enchanted us as children—one that adults often forget. Read these stories about nonhuman animals and reawaken your perspective on life."—**Elizabeth N. Goodenough,** University of Michigan, and editor, *Secret Spaces of Childhood*

"Beautifully written, William Crain's personal account of rescuing farmed animals and caring for them at his sanctuary contains inspiring insights into the minds of animals. These moving stories will brighten your day and leave you filled with wonder."—**Maya Gottfried**, author, *Our Farm*, *Good Dog*, and *Vegan Love*

"This wonderful book gives an in-depth view of what goes into creating an animal sanctuary. But more importantly, the stories of the animals provide a glimpse into their lives as unique individuals, deserving of respect, kindness, and above all, love."—**Dr. Joanne Kong**, editor of *Vegan Voices: Essays by Inspiring Changemakers*

"William Crain's new book is more than a collection of animal stories; it is a gentle and moving reflection of narratives about the very marvel of *aliveness* itself. In story after story, we are introduced to lives, different from our own, yet always, at the center of them all, our common and instinctive need to survive—and to communicate and speak with each other. Indeed, the very farm, where all these stories take place, is truly a habitat of just such a shared aliveness, an expression, surely, of the wonder and beauty of each of us, *being* here."—**Richard Lewis**, Touchstone Center for Children and author, *Living by Wonder*

ANIMAL STORIES

LIVES AT A FARM SANCTUARY

BY WILLIAM CRAIN

Lantern Publishing & Media • Woodstock & Brooklyn, NY

2024
Lantern Publishing & Media
PO Box 1350
Woodstock, NY 12498
www.lanternpm.org

Copyediting and design by Pauline Lafosse

Printed in the United States of America

Library of Congress Cataloging-in-Publication Data

Names: Crain, William C., 1943- author.
Title: Animal stories : lives at a farm sanctuary / William Crain.
Description: Woodstock, NY : Lantern Publishing & Media, [2024] | Includes
    bibliographical references.
Identifiers: LCCN 2023028829 (print) | LCCN 2023028830 (ebook) | ISBN
    9781590567203 (paperback) | ISBN 9781590567210 (epub)
Subjects: LCSH: Animal rescue—United States—Anecdotes. | Livestock—
    United States—Anecdotes. | Animal sanctuaries—United States—
    Anecdotes. | Animal rights. | BISAC: NATURE / Animal Rights |
    TECHNOLOGY & ENGINEERING / Agriculture / Animal Husbandry
Classification: LCC HV4708 .C695 2024 (print) | LCC HV4708 (ebook) |
    DDC 636.08/320973—dc23/eng/20231212
LC record available at https://lccn.loc.gov/2023028829
LC ebook record available at https://lccn.loc.gov/2023028830

To Karen Davis,
who fought so valiantly in defense
of our animal relatives.

To the Reader

Almost all the chapters follow the order in which the animals came to our farm. The exception is chapter 26, the story of Emma, a turkey who lived an extraordinarily long life. It felt natural to place her story near the end.

Throughout the book, I use the term "animals," rather than the more precise "nonhuman animals," for the sake of simplicity.

Our staff members have contributed greatly to the farm. The three who most frequently appear in the book are:

- Stacy, our first caretaker
- Joy, our longest serving head caretaker
- Chris, who began volunteering as a twelve-year-old and worked as a paid staff member from ages sixteen to twenty-one.

CONTENTS

1

STARTING A FARM SANCTUARY

When my wife Ellen and I entered our 60s, we began a new project. We renovated a run-down farm in upstate New York and turned it into Safe Haven Farm Sanctuary. The sanctuary provides a lifelong home to farm animals rescued from slaughter and abuse. We have been taking in animals for the past 16 years and currently have 170 animals, including goats, sheep, chickens, turkeys, pigs, donkeys, cows, and horses. In this book, I will tell you about some of these animals. First, though, I should describe how we began the farm sanctuary.

MAKING THE DECISION

Ellen and I were busy with our professional lives. Ellen was the director of the pediatric emergency room at Jacobi Medical Center in the Bronx. I was a psychology professor at The City College of New York. Our occupations demanded a great deal of time. We weren't looking to add a major new activity.

But another force was at work within us. Over the years, we increasingly found our hearts going out to animals; we worried about their plight.

Ellen was reading a great deal about factory farms, which provide most of the meat sold in the United States. Factory farms don't fit the popular image of a farm. They do not have pastures in which animals happily graze, nor do they have barns in which the animals eat hay and bed down for the night. Instead, factory farms consist of huge concrete buildings in which chickens, pigs, and other animals are crammed into spaces so small they can barely move. Most of the animals are forced to stand on wire or concrete. They will never experience sunlight, breezes, soil, or grass.

Ellen wanted to do something to help these animals and read about the impressive work of the Farm Sanctuary in Watkins Glenn, New York. This sanctuary, like the others that have followed in its footsteps, cares for animals who somehow escape factory farms. For example, a cow, chicken, or pig might occasionally escape a slaughterhouse. Or some animals might get free of their crates when a transport truck gets into an accident. Passersby see the animals wandering loose, want to find them a safe home, and call the sanctuary to see if it can take them.

Farm sanctuaries can only save a tiny fraction of the 10 billion land animals that are slaughtered for meat each year in the United States. But the sanctuaries can also host visitors and publish articles about their animals, making the public more familiar with the animals that are being killed. People might then give more thought to their food choices.

Ellen was initially more enthusiastic about creating a farm sanctuary than I was. I agreed that farm sanctuaries are important, but my preferred way of working for a cause was political action. I frequently spoke at public hearings and participated in protest rallies. I had even engaged in civil disobedience in an effort to protect trees and black bears. I wondered if I would have the patience to care for numerous farm animals.

But the more I thought about starting a farm sanctuary, the more I liked the idea. I would get a chance to learn about nonhuman animals, who have provided my profession—psychology—with insights into human behavior. For example, psychologists have learned about human toddlers' attachment to parents by examining attachment in other species. In recent decades, animal research has lost some of its popularity in psychology, but it remains important. Perhaps, I thought, a first-hand knowledge of animals could contribute to my teaching and academic writing. This possibility began to excite me, and I agreed with Ellen's wish to establish a farm sanctuary.

GETTING THE FARM READY

We purchased the farm in 2006 and looked for a contractor to renovate the buildings on the property. These structures—a small house, cottage, and barn—were in such dilapidated condition that the first three contractors told us to tear them all down and start over. But among the shambles was some fine wood craftsmanship, so we kept searching until we found a contractor who wanted to restore them.

Construction took two years. Once it was completed, we named the farm Safe Haven Farm Sanctuary and got ready to open it to animals. Because Ellen and I couldn't live full-time at the farm yet, we hired a caretaker, a young woman named Stacy. Stacy, who had been living in Connecticut, drove to the farm with her beautiful mare and a chicken who had been abandoned in a pasture with a broken leg. The chicken, a hen, was our first rescued animal.

The next group of animals came from a live meat market in the Bronx. My experience at the market was one of the worst of my life.

We restored a run-down farm

2

THE LIVE MEAT MARKET

Live meat markets allow customers to walk in and select the kind and size of the animal they want slaughtered and butchered. This is done on the premises, so customers can be certain their meat is fresh. In most of these markets, the majority of the animals are chickens, which are usually crammed on top of one another. Many of the markets also sell ducks, guinea hens, rabbits, goats, and sheep.

Ellen and I had driven by a market in the Bronx several times, and we had often seen a large sheep standing near the front door. Sometimes we saw one of the goats. Occasionally, people purchase an animal from a live meat market to save their life. We later learned that most farm sanctuaries oppose this; the purchases, they point out, support the markets' business. We came to see their point. But at the time, the sight of the animals made us want to do something. So when our farm sanctuary was ready to house animals in February of 2008, we decided to save a few from this market. In particular, I had my mind on the sheep by the door. We also decided we would purchase two goats.

Ellen and I rented a truck, put hay in the back, and drove to the meat market. It was a chilly Sunday morning. When we arrived, we were joined by our farm's first caretaker, Stacy, and her friend Tom. We walked into the store, talked to the manager, and purchased the sheep we had seen by the door. We also bought a second sheep—a recent arrival who was the only other sheep on the premises. The two sheep frantically tried to escape being grabbed by the workers, but they were caught within two or three minutes.

When I turned my attention to the goats, I started feeling upset—even a little sick inside. How could we decide whom to save? Saving two meant determining that the others would die. But the manager didn't ask us to choose. He just told his workers to capture two goats. As soon as the workers stepped into the goat's room, all of them, like the sheep, ran for their lives. The goats were quicker and more difficult to catch. It took the workers about fifteen minutes to grab two of them. So we didn't make any selection; our two goats were simply the two the workers could capture.

Once a worker caught a sheep or goat, he held the animal by the feet and flung the animal upside down. He then dragged the sheep or goat about thirty feet across the cement floor. When he reached the large scale for weighing, he flung the animal onto it. If the animal was very heavy, another worker helped throw the animal onto the device. We pleaded with the workers to be gentler, but they ignored us. We paid the manager, put makeshift leashes on the animals, and with some pushing from behind, guided them into the back of the truck. They were all trembling in fear.

On the trip back to the farm, Ellen, Stacy, and Tom drove a car, and I drove the truck with the goats and sheep. The trip would ordinarily take an hour and a half, but I lost contact with the cars and took a wrong turn. For another half hour, I couldn't figure out how to get back on the correct highway. I soon began to worry about the animals' health. *The truck doors are all closed,* I thought. *If this trip takes longer than I planned, will the animals get enough air?* I knew nothing about such transports. Should I pull the truck over and open its back door to give the animals some air? Or might they jump out and run onto the highway? Finally, I did stop and peek in. They were huddled tightly together but breathing okay.

After three hours of driving, I steered the truck up to the barn. Ellen, Stacy, and Tom were waiting. I sensed that they were irritated by my delay, but they didn't say much because we all had to focus on getting the animals out of the truck and into the barn. All the sheep and goats were scared to death. Pointing to the smaller goat, Tom said, "This one is trembling something awful!" Tom tried to calm her by stroking her back, but to no avail. He told her, "You're scared now, but you don't know how lucky you are." That night, Stacy named the animals and called this small goat "Mattie."

We kept the sheep and goats in a quarantine stall for four weeks, while we had them tested and treated for parasites and other illnesses. During this period, Ellen and I had to work in the city, but we traveled to the farm several days a week. I spent most of my time sitting quietly in a corner of their stall, hoping that if I were unobtrusive, they would get used to me and become less fearful of humans. Sometimes, I sang in a low voice. This approach seemed to work, but only slightly.

When we finally let the goats and sheep into the pasture, their first act was to inspect the fences. They walked to each section of fence, sniffing and examining it. All the while, the farm's large mare was watching from outside the fence. Then, when the sheep and goats finished their inspection and drifted to the center of the pasture, the mare began running, jumping, and snorting. To me, she seemed to be saying, "Come on! Play!" The goats and sheep spent about thirty seconds running and jumping, too, but then stopped. They weren't comfortable enough to continue.

It would take a few more weeks for our first sheep and goats to feel at ease on the farm. It would take much longer—even years—before all of them would interact spontaneously with humans. But in the meantime, their lives were filled with adventures.

A live meat market

3

MATTIE

Our Small Goat Becomes a Mother

Mattie was rapidly gaining weight. We wondered if she could be pregnant. Our vet examined her, but he couldn't tell. Then, after four months, Mattie went into labor. But the baby was turned the wrong way. Stacy, our farm's caretaker, took charge and turned the baby around. She said the baby then came out so rapidly it was "almost like a cannon boom." She named the baby, a boy, Boomer.

Baby Boomer was full of energy. By seven days of age, he was sprinting back and forth in the aisle of the barn, just for the fun of it. When he was ten days old, he ran out to the pasture in the morning and tried to climb up a rock that was about one and a half feet high. Just as he neared the top he slid down backwards, landing with a thud. He climbed back up and jumped down—backwards—and this time he landed perfectly.

Boomer climbed up and jumped down several more times. Sometimes he leapt forward, sometimes backwards, and he added new spins while in the air. He looked like a platform diver experimenting with new stunts.

All the while, Mattie looked on from a distance, but she didn't intervene. According to the great Dutch animal researcher Niko Tinbergen, this kind of unobtrusive presence is typical of parents in many species. The parents constantly keep an eye on their young, but they act only when necessary.

As Boomer grew up, he calmed down somewhat, but not completely. He still loved to play. Mattie, too, was rambunctious. Both Mattie and Boomer were often up to mischief. They found ways to get into the chicken coops to eat the chicken food, and they figured out how to squeeze through fences to gain access to tasty new plants and trees. Like our other goats, they were enjoying life on the farm.

Then, when Boomer was two years old, he suddenly didn't seem himself. He was sluggish. We called our vet, who examined Boomer at the farm and told us to test Boomer and the other goats for parasites. The results revealed that Boomer was suffering from severe anemia; a parasite was sucking the blood from his stomach. Mattie had the same parasite, but her case was not as severe.

Our vet said Boomer's life was in imminent danger and recommended that we drive Boomer to the Tufts Veterinary School in Massachusetts, three hours away from our farm. Ellen and a new caretaker, Karen, put Boomer in the backseat of our Honda Civic. They put Mattie in the car, too. Mattie could have been treated at the farm, but we didn't want Boomer or Mattie to be left alone.

When Boomer reached Tufts, the staff was waiting for him. Once he was out of the car, he tried to walk, but he staggered, so the staff lifted him onto a stretcher.

Although Boomer was weak, he wouldn't lie down on the stretcher. And although he sometimes wobbled, he remained standing as the staff carried him down a long hallway to the examining room. Our vet later told us this behavior is natural in goats. Whenever they fear danger, they stand. Having lived millions of years in the presence of predators, their instincts tell them to be ready to run.

Ellen and Karen watched Boomer go, while Karen held Mattie on a leash. Karen held it tight. She knew that Mattie loved to explore and at this hospital there were many things to attract her—hospital equipment, stacks of hay, and open doors. Mattie could get into considerable trouble.

Then Ellen told Karen, "Let Mattie go free."

Karen was perplexed. It seemed like a crazy idea. But Ellen was the boss, so she followed Ellen's order. Once off the leash, Mattie rushed to catch up with Boomer, and then walked directly behind the stretcher for the rest of the long trip down the hall. For Ellen and Karen, it was a touching sight. Mattie could go wherever she wanted, but all she cared about was being with her son.

The Tufts staff did a great job. They had to repair Boomer's stomach, but he recovered. Mattie recovered, too.

Boomer grew to be much larger than Mattie, so much so that visitors often have difficulty believing that big Boomer is her son. But even as grown-ups, they usually rest snuggled up together.

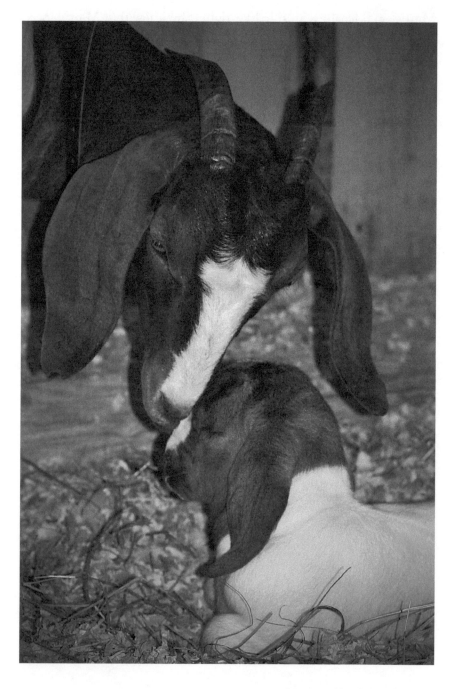

Mattie and her baby, Boomer

4

SPRINKLES

A Leader Among Our Sheep

Before we opened the farm sanctuary. Ellen and I sometimes drove by a live meat market where we saw Sprinkles. She was a large sheep who was typically looking out of an opening in a door. I initially thought she was interested in events on the street. But as I observed her more often, I saw a steadiness to her gaze. She seemed to focus on a spot far in the distance. She also appeared calm. I was puzzled. How could she be calm? Surely, she knew she was awaiting slaughter.

Even though she appeared serene, we wanted to get her out of there. When we went to the market to bring animals to our newly opened farm, we purchased her, as well as a second, smaller sheep on the premises. Once we got back to our farm, our caretaker named the large sheep Sprinkles and the smaller sheep Angel. Sprinkles and Angel lived together in the same stall.

We didn't know Sprinkles' age or background. The only clue to her past was her ears. They were missing pieces at the tips, signs of frostbite that probably came from enduring extreme cold on farmland.

Over the years, we adopted several more sheep. All considered Sprinkles to be the leader. Whenever the sheep moved about together, Sprinkles led the way.

Sheep are often timid or "sheepish" around humans. Sprinkles didn't fit this description. On several occasions I saw Sprinkles walking toward something when a human moved into her path. Sprinkles just kept walking, and the person had to move.

But this doesn't mean that Sprinkles disliked humans. She liked receiving a pat on the head from a person she knew well. Sprinkles simply didn't show humans any special deference.

After seven years with us, Sprinkles suddenly lost weight. We drove her to the Tufts Veterinary Hospital in Massachusetts for tests and treatment (taking Angel along so neither sheep would be alone). But the doctors couldn't reverse Sprinkles' illness. They believed she was a very old sheep whose immune system had finally given out.

We brought the two sheep back to our farm and, following the hospital's instructions, we gave Sprinkles intravenous liquids and medication in her barn. But after a few days, Sprinkles died.

Sprinkles died in the daytime, and Ellen and I wondered where to put Angel for the night. Would it be best to keep Angel in the same stall she had shared with Sprinkles? We decided to leave the decision to Angel. When we began putting the animals to bed, Angel walked into the barn, looked into her stall, and saw that Sprinkles wasn't there. She then quietly turned around and joined some other sheep for the night. Angel didn't want to be in her former stall without her friend. And she never went back.

Our farm has a small cemetery for the animals who pass away, and we hold ceremonies for them. Many tears were shed when we laid Sprinkles into the ground.

When I think about Sprinkles' life with us, my mind often turns to the first day I saw her—when she calmly looked out of the meat market to a point in the distance. Since then, I have occasionally come across reports of similar gazing in other species. I also have seen similar behavior in some animals on our farm, especially the goats.

The poet Rainer Maria Rilke wrote about this gazing into space. He said that whereas humans look out and see individual objects, animals can look into "the Open" and lose themselves in the timeless vastness. Rilke said they feel themselves to be part of the eternal. If Rilke was correct, Sprinkles, awaiting death in the meat market, might have appeared serene because she experienced herself as part of something that goes on forever.

We can only speculate about such matters, of course. And, in any case, Sprinkles didn't die in the meat market. She enjoyed seven more years of life at our farm sanctuary. Those of us who knew Sprinkles are thankful that she lived with us, and we still miss her.

Sprinkles

5

The City Girls

Chickens Respond to Freedom

Our farm sanctuary still struggles with how much freedom to give the animals. We don't like them to be confined, but we also must keep them safe from hawks, weasels, and other predators. We especially worry about our chickens, who are very vulnerable to predation.

We initially erred on the side of safety. We kept all the chickens in enclosed spaces—in barns with attached aviaries. After a while, we began to experiment with allowing the chickens time in open pastures.

Our first "experiment in freedom" was with a group of 19 chickens—mostly hens—who came from a low-income section of Manhattan. A homeless man felt sorry for the chickens in a live meat market. Whenever he got enough money together, he purchased a chicken and lifted her over a fence into a vacant lot. Neighbors often tossed vegetables into the lot for the chickens to eat, which kept them alive.

But some neighbors worried about the chickens' health and called the ASPCA, which found that the chickens were malnourished and lacked protection from the elements. The ASPCA was happy to learn that we would adopt them, and four ASPCA volunteers drove them to our farm. We called their group "The City Girls."

There were many signs that the chickens had endured hard times. They were scrawny and missing many feathers. They also were missing the fronts of their beaks, a sign that they had initially been raised on factory farms. Factory farms pack chickens so tightly together that they fight, which means they damage the owners' "products." If they would give the chickens more space, the fighting would stop. But the owners make more money by crowding them together, so they reduce the damage by clipping the birds' beaks, usually without anesthesia.

We housed the chickens in a barn. Attached was a small, mostly dirt yard. After a few weeks, the chickens had gained weight and were in good physical health. But they were constantly irritable. They walked around aimlessly and made angry clucking sounds. They became upset if they saw any human visitors.

One sunny spring morning, Ellen and I decided to let them out of the barn into the large adjoining pasture. I opened the barn door, and a few tentatively ventured out. The rest followed. I sat on a rock to keep an eye on them. Within a few minutes, they were eagerly foraging, scratching the ground to see what they could find. They especially liked digging under leaves. After an hour or so, I steered them back inside.

Back in the barn, their behavior completely changed. All their irritability vanished. As they walked about, they uttered lovely, contented sounds.

Later that day, two people visited the chickens, and whereas the chickens had previously hated all visitors, the chickens weren't disturbed in the least. Freedom had transformed them.

Since then, we have given our chickens time in open areas whenever the weather permits and we can keep an eye on them, and they have always seemed happy.

I would add that I don't think it was just freedom, in and of itself, that altered our chickens' emotions. I have observed so-called "free range chickens" who are free, in the sense that they have space to move, but can't really forage. They can only move on a hard dirt surface. And they don't seem happy. They don't emit any sounds of contentment. It's not just freedom, then, but the freedom to engage in natural activities like foraging that produces happiness.

The City Girls

6

BURDOCK

Our Bantam Rooster

One day we received an unusual phone call. It was from the director of our town's recreation department. She said that a petting zoo had brought several farm animals to the town's community day. But when the event ended, the zoo's workers couldn't catch their bantam rooster, and they left him in the parking lot. The director was worried because there were coyotes and other predators in the area. She also was concerned about the danger posed by cars. She wondered if we could send anyone to capture the rooster and give him a home.

Two of our caretakers, a young woman and young man, were eager to try. When they got to the parking lot, they saw the little rooster, a very colorful fellow who was only about half the size of a typical hen. They soon discovered that he was also a fast runner and a good flier—and extremely difficult to catch. Whenever they thought they had him cornered, he escaped their grasp. Finally, after two hours, he ran into a hollow log in the nearby woods, and they were able to grab him.

When the caretakers returned with the little rooster in a crate, they told us they wanted to name him Burdock, a name which they thought fit his handsome appearance. Ellen and I agreed with the name.

Next, we all thought about where to house him. The obvious choice was our aviary, an outdoor area enclosed by a wire screen attached to a coop. But we guessed that Burdock was so wild that he would go berserk if he was confined in any way. After considerable discussion, we decided to let him roam freely. We were nervous about the decision because we didn't know if he could become the target of a predator, but it seemed to be the only way he would be happy.

During the daytime hours, Burdock wandered here and there, climbing on fences and roofs, and crowing whenever the mood struck him. If another rooster tried to push him aside, Burdock always fought back, even when the other rooster was three times his size.

We were happy to see that Burdock wanted to sleep inside the barn. There, he was at least safe from weasels, coyotes, and predators who generally hunt at night. Burdock roosted on the highest rafters, usually out of sight. In fact, when we conducted our count of all the farm's animals at dusk, we usually needed a flashlight to find him. Then, after a few months, Burdock's daytime roaming suddenly stopped. The change occurred when we adopted Sweetie, an elderly hen who had been found in a vacant lot in Brooklyn.

Sweetie had a severe limp, and our vet said there was no remedy for it. The vet explained that Sweetie had all the signs of a hen who had been raised for meat on a factory farm. These hens are bred to gain so much weight that their legs have trouble supporting them and frequently give out. Soon, Sweetie couldn't walk at all. Each morning, we carried her from her indoor cubby to the grass in the aviary so she could enjoy the sun and breezes.

To our amazement, Burdock began spending his daytime hours quietly sitting near her, just outside the aviary fence. And one day, when we opened the aviary door to go in, Burdock walked inside and sat down next to her. After that, he rarely left Sweetie's side. He seemed to want this elderly hen to have his companionship.

A few weeks later, Sweetie passed away. Burdock then began paying attention to the other hens in the aviary. Although he frequently wandered outside the aviary, he always went inside at dusk and called any hens who were still outside to come in. He wanted everyone to be safe for the night.

Burdock never lost his wild side. He never felt comfortable around humans, and he was always ready for a fight with other roosters. But he demonstrated that even a wild, freedom-loving individual is capable of devotion and caring.

Burdock

7

KATIE

An Exceptionally Caring Hen

Katie was brought to us by a young couple from Brooklyn. They had heard that chickens make good pets and had purchased her from a live meat market where she was awaiting slaughter. The couple named her Katie and brought her to their apartment, only for their landlord to say, "No chickens allowed." And so, the couple looked for a place where Katie would be happy and safe, and they found us.

We learned that Katie had a very caring nature. The first hint of this was provided by a little partridge we named Cleo. Cleo found her way to us from the hunting club over the hill. She had escaped the guns and was wandering about our farm, seeking refuge. She entered Katie's aviary and soon looked to her for protection. Whenever Cleo became frightened, she ran over and snuggled under Katie's wing. Cleo sensed something in Katie that made her feel safe.

One afternoon, Cleo wandered outside the aviary and eagerly explored the surroundings. I enjoyed watching her, but when it started getting dark, I wanted her to go back inside the aviary, where she would be protected for the night. But I couldn't coax her in. The more I tried, the more she resisted. I became very worried because many predators come out at night—like owls, foxes, weasels, and coyotes.

We were losing hope when Katie walked out of the aviary and went directly over to Cleo. Katie looked at Cleo for a moment, then turned around and returned to the aviary. And Cleo followed Katie back in.

We cannot know what exactly transpired between Katie and Cleo, or what Katie might have been thinking. But it seemed to us that Katie acted like a mother guiding her little one back into the home.

Another surprising event took place when Ellen and I had to give medicine to Katie and Burdock, our bantam rooster, who lived in the same aviary.

We were able to give the medicine to Katie without much difficulty, but Burdock put up a great fuss. He squawked and flew about. Every time we almost got a hold of him, he darted away. Finally, we had him cornered. He had run out of escape routes. Then Katie ran over and put her body next to his, blocking our access to him. We were sure she was trying to protect her friend. We were eventually able to get around Katie and give Burdock the medicine, although I cannot remember how we did it. What will always stick in my mind was how Katie tried to intervene on Burdock's behalf.

One time, Katie even seemed to demonstrate caring feelings toward me. I was upset by news of a relative's illness, and as I went about the farm chores, I felt a bit like crying. Katie walked up to me (which she had never done before) and looked into my eyes.

To some people, it will seem to be a stretch to say that Katie was concerned about me. They will think that I am anthropomorphizing, attributing a human emotion to a nonhuman animal.

But I had the strong feeling that Katie sensed my pain and was concerned.

After three years with us, Katie began to weaken. She continued to forage outside the aviary, but she spent more and more time resting inside it. One afternoon, she sat there with her eyes closed. She was slipping away.

By this time, the aviary also housed another chicken hen, three partridges, and Burdock. During her last two hours, all the birds stayed by her side. Burdock remained closest of all.

When Katie died, Burdock rose and flapped his wings with great vigor. Then he crouched down very low, rose higher than ever, and sent out a piercing rooster's cry.

Someone special was gone.

Burdock and a partridge with Katie in her last hours

8

THE TURKEYS AND THE GIRL SCOUTS

During our sanctuary's second fall, four female baby turkeys were dropped off at our farm. They were left anonymously, so we had no background information. But we were pretty sure that they began life on a factory farm. For one thing, their feathers were all white. Factory farms breed turkeys to be all-white to prevent blotches of pigment on customers' meat.

In addition, our turkeys, like most of our chickens, were missing the front ends of their beaks. Factory farms cut them off to make it harder for them to tear each other's flesh when they fight—fighting that stems from overcrowding. Our turkeys were also missing the ends of their toes, which had been removed to further reduce the damage from fighting.

The young turkeys slept in the barn at night and enthusiastically explored the farm during the day. They frequently traveled as a group, and whether they ventured together or alone, they were constantly on the move, chirping away.

One day a troop of Girl Scouts visited our farm. The girls— nine in all—brought several useful things for the farm and held a ceremony in the barn. They stood quietly in a circle while they took turns standing in the middle and reading pledges. They vowed to always cherish animals and protect their right to a full life. One pledge promised to console dying animals and to "ask the angels to gather them in their arms." The girls' statements were very solemn.

Soon after the Scouts began reading their pledges, our four turkeys, who until that point in their young lives had spent every waking moment noisily milling about, joined the circle. Each turkey moved to a separate section of it. They sat in perfect stillness, with their eyes fixed on the reader. It was as if the turkeys were listening to the pledges and were moved by them. Those of us who worked at the farm were astonished.

The turkeys could not have understood the Scouts' words, of course, but they must have felt the Scouts' inward quietness. I suspect there was a spiritual element—an attitude of quiet reverence—in the Scouts' ceremony that the turkeys responded to.

Our four turkeys

9

DUCKY

A Turkey Who Disliked Humans

We soon learned that each of our four turkeys had a unique personality. Ducky often seemed agitated, as if there was something she wanted but couldn't find.

Ducky also disliked humans, and one person above all. This woman, a staff member, never mistreated Ducky. She merely carried out her tasks in a perfunctory manner. She cleaned the barn floor and put out the turkeys' food as if the jobs were all that mattered, ignoring the turkeys themselves. Ducky hated her and frequently tried to peck her legs.

At the same time, Ducky displayed concern for the other turkeys, as well as her intelligence. One day, when the turkeys were still young and light enough to fly, our turkey named Sadie flew over a pasture fence. Ducky became upset by the separation and began emitting distress calls.

At that moment, Grant, another staff member, was returning from a swim in our pond and went to see what was wrong. When he approached Ducky, she looked directly at him and made louder calls. He was sure that Ducky wanted him to retrieve Sadie. But Grant had sandals on his feet, which would make climbing the fence difficult, so he headed toward the cottage to get his boots.

When Grant started walking *away* from Ducky, she called out with even greater urgency. Grant later told me, "It was as if she was saying, 'Where in the devil are you going? I need you to stay here and get Sadie.'"

Grant was so affected by Ducky's desperation that he decided to climb the fence in sandals, and he succeeded in retrieving Sadie. Only then did Ducky calm down.

Turkeys are sometimes believed to be stupid. But Ducky demonstrated a knowledge of means and ends. She knew that Grant was the means of returning Sadie. Many scientists consider means/ends behavior to be a distinguishing mark of intelligence.

Although Ducky never seemed to like humans, she did something in her last year of life that was completely out of character:

A group of Buddhists, fifty in all, visited our farm. They were all initially from China, and only one spoke English. After touring the farm and eating lunch, they asked if they could perform a blessing ceremony. The head monk, who was revered by all, picked some leaves from a nearby evergreen and made "compassion water." He then led all the Buddhists in single file as they walked throughout the farm. He led chants and sprinkled water here and there. The ceremony took half an hour.

After the ceremony, they were chatting together outside the barn when Ducky, who by then had such bad arthritis that she could barely walk, hobbled out. She slowly approached the monk and looked directly at him for several seconds. After that, she was interrupted by others who jumped in to take photos of the encounter.

Why did Ducky walk over to the monk? I have no explanation. Buddhists might think that Ducky knew the monk or his ancestors in a past life. As a psychologist trained in Western science, I never had put stock in such ideas. But this event was so unusual I began to wonder.

Ducky

10

THE DUCKS WHO ESCAPED THE GUNS

One winter evening, our neighbor Jim phoned me about twelve ducks he had seen next to his house. He said they had sought protection from a snowstorm. "But only one is left," Jim said. "The others wandered off, and I think they froze or starved to death. I've seen a few frozen in the snow."

Jim asked us if we would take in the remaining duck, and we agreed. He was a male mallard. We expanded an aviary for him and installed a small pond in which he could swim. But the problem of the ducks was a big one, and it couldn't be handled with aviaries.

Jim was sure the ducks had escaped a hunting club a mile away. When we visited it, we heard dogs howling in pens and saw many bags of dead birds on the grounds. Soon, an employee approached us, told us that visitors weren't welcome, and ordered us to leave.

But we continued to gather information. We learned that the club charges high membership fees for the privilege of shooting ducks, pheasants, and partridges. Because the birds were raised in indoor factory farms, each bird's release is probably the first time the animal sees the light of day. The birds don't have time to orient themselves to their surroundings and are easy targets. Miraculously, some birds escape, and people see them wandering in our area.

Jim, who had lived on our road for many years, told us that he often saw a few mallard ducks on our pond in the fall. But when the pond froze during the winter, the ducks disappeared. He believed that predators, like foxes and weasels, got them. When ducks have access to water, he explained, they have a good chance of avoiding predators; ducks can easily out-swim them. But when a pond freezes, the ducks have no water to escape to, no safe port. Ordinarily, wild mallards will migrate south before their ponds freeze, but these ducks didn't know how.

Before the next winter arrived, we tried to help the ducks by installing two bubblers in our pond. Bubblers are small machines that stir up water. And we were very happy to see that the bubblers worked; they prevented an area of the pond from freezing and the ducks had a place to swim.

Only a few ducks who came to our farm found their own way to the pond. Most wandered alone in a confused manner, and Ellen or I had to steer the duck toward the water. We walked behind the bird, who waddled forward to avoid us. As soon as the ducks saw the others on the pond, they became excited and joined them. We worried that the newcomers might not be welcome, but they always were.

When there were about fifteen ducks on the pond, I began giving them some corn and grain in the mornings. The first time I put the food out, they ate it eagerly. But one duck stayed back. He was a male who appeared to be weak, and he flapped one wing as if injured. I thought, *Oh, no. He's sick or hurt. But he's far from the shore. How will I ever get him to take him to a vet?*

Later that day, I walked down to the pond to get a better look at him. Surprisingly, he was large and looked healthy. He had exceptionally bright coloring.

When I put out the food the next morning, he again stayed back, without any sign of injury. And this time, after the other ducks had eaten for a few minutes, he swam to the shore where they were feeding. He looked at them, swam back toward the center of the pond, and the others followed him. He was the leader!

Over the next few days, the same thing happened every time: While the others ate on shore, he stayed back. He was apparently standing guard until he decided it was time to lead them back to the safety of the water.

Why did he appear to be injured the first time? I believe he was engaging in a diversionary tactic, as birds sometimes do when predators are near their young. He was trying to draw my attention to him, pretending he was vulnerable prey, so I wouldn't attack the others.

This duck's leadership marked the beginning of social organization. After that, I was impressed by how the ducks acted as an organized group. The most impressive example of their group behavior occurred when two geese came to our pond.

The geese were much larger than the ducks and soon began to dominate the feedings. When ducks came near the corn and grain, the geese chased them away. I felt sorry for the ducks and began to reconsider my practice of giving out food.

Then one morning several ducks huddled in the middle of the pond. Suddenly, they flew, *en masse*, toward the geese. They flew with their heads low, in attack mode, like an airplane squadron. The startled geese flew away.

When the geese returned the next day, a truce developed. The ducks and geese shared the food.

Later I was re-reading Konrad Lorenz's book, *On Aggression*, and learned that it is not uncommon for prey animals (which include ducks) to form groups and attack even fearsome predators.

It is called "mobbing." You might have seen birds flock together and harass a cat.

In any case, the ducks behaved as an organized unit. They were no longer the disoriented individuals who helplessly wandered about our farm.

The hunting club (top) and ducks on our pond in winter

11

SYLVIA

A Jersey Girl

Sylvia came to us from a small town in central New Jersey. There, she was among a group of chickens who roosted in trees overlooking a tent city. When the people had their meals, they regularly tossed some of their vegetables to the chickens.

But the town's officials didn't want any homeless people in their community and drove them out. When two animal lovers heard about this, they worried that the chickens might not survive without the human handouts. The two people decided to rescue the chickens, and they captured as many as they could—eight of the twenty they saw. We agreed to adopt the chickens, who were all hens. We called them "the Jersey Girls."

They were different from our domestic hens. They were slim and athletic, and they flew with the grace of other birds. And they didn't want to sleep in a henhouse. They wanted to continue spending their night in trees.

But we were nervous. At the time, a fisher cat—a fierce, tree-climbing member of the weasel family—had been sighted in our area. We didn't want to take any chances while this predator was around. We therefore coaxed the Jersey Girls into the henhouse at night, although it was difficult. They didn't like taking directions from humans.

Sylvia was among the most independent. Even after five years with us, she sometimes liked to wander out of our sight and to sleep high up in the barn rafters.

One evening, Ellen and I were tucking in the animals for the night but couldn't find her. We looked in all the areas she frequented, but she was nowhere in sight. As darkness fell, I walked to the woods with a flashlight. I didn't expect to find her, but I thought I'd give it a try.

I didn't see her. Giving up, I began walking back to the barn, when I heard a chicken squawking in the woods. I went back and saw Sylvia high in a tree. She had wanted me to find her.

Ellen and I carried Sylvia down from the tree and back to her henhouse. She went willingly.

I don't know what made Sylvia fly up in the tree. Perhaps a predator, like a fox, frightened her. In any case, what stood out was that she didn't remain silent, as she would have in the past. She squawked to get my attention. It felt to me like she was saying, "I trust your care. I didn't for a long time, but I do now."

Sylvia

12

E.J.

An Aggressive Duck

E.J. was a male Muscovy duck who came to us from Long Island, where a family found him swimming in their backyard pool. How he got there, nobody knew.

The family called a local wildlife rehabilitation center, which temporarily cared for him until we took him in. The center named him E.J. in honor of a duck it rescued in the past.

A clue to E.J.'s history was that his upper bill was missing, indicating he was raised for meat on a factory farm. As with chickens and turkeys, these farms house ducks in overcrowded conditions that cause frequent fighting; the animals tear at each other's flesh. The farmers reduce the damage to their "products" by cutting off the ducks' upper bills. Because E.J. had no top bill, he couldn't forage for plants or insects, and we soaked his food pellets in water to make it easier for him to eat.

When E.J. arrived at our sanctuary, he was very aggressive. Even though he was missing his upper bill, he constantly grabbed onto our pant legs with his mouth and often wouldn't let go for a few minutes. If we shook the leg to try to get free, he grabbed it with his sharp claws and slapped us with his wings.

Most people didn't want to go near him. But we had no choice; he ventured into places where he caused trouble, and we had to fetch him. For example, he sometimes went into a barn where he fought with the turkeys or snuck into chickens' aviaries to eat their food. We had to go after him and do our best to shoo him out.

For several months, E.J. seemed unmanageable. But two people began to change his ways. One was Devon, a young woman who volunteered on Saturdays. Other staff observed that E.J. became very obedient with her, coming whenever she called.

I asked Devon, "What's your secret?" She said, "I never tried to get him to behave. I just showed him I loved him. At first, he grabbed my pants, and I was afraid of him. But I began talking to him in a soft voice, telling him what a handsome fellow he is and that he is my boyfriend. I petted him, and he wagged his tail feathers. Now he's always nice and sometimes comes up to me."

The other person who significantly altered E.J.'s behavior was Ellen, my wife.

Ellen's view of his behavior struck most of us as odd. Ellen said, "When E.J. grabs hold of people's pants, he isn't trying to be mean. It's his way of connecting with people. He's seeking attachment."

I still have my doubts about Ellen's interpretation, but because she saw his behavior as something he needed to do, she demonstrated great patience with it. When E.J. grabbed hold of her pants, she never tried to shake him loose. Instead, she allowed him to decide when to let go. It was quite a sight to see Ellen walking about, performing her chores, with E.J. hanging onto her leg. Sometimes it was a minute or two before he released his grip.

To my surprise, Ellen's patience had a positive effect. After a while, he completely stopped grabbing her pants. He apparently concluded that Ellen would never hurt him and began to enjoy being in her presence. And even when he ventured into areas where he could cause trouble, Ellen only had to call his name and he waddled over to her.

The changes that Devon and Ellen produced gradually transferred to other people. E.J. became calmer and more obedient with everyone.

Obedience is, of course, a major goal of pet trainers, who typically emphasize the need to let the animal know that a human is the boss. The trainer often instructs the pet's owner to assume the role of an alpha member of an animal's group. Obedience is also of great importance to most parents and those who work with human children. They, too, try to take charge and exert control.

E.J.'s obedience came about in a different way. Devon and Ellen won him over with love and patience—an approach that should be tried more often.

E.J.

13

Chicken Little and Her Two Children

One day my wife, Ellen, greeted me with the words, "Congratulations, you're a new grandfather." I was puzzled. I didn't know of any new births in our family. Then she explained that one of our chickens, a small hen we called Chicken Little, had hidden two eggs under a porch and hatched them. My new grandchildren were two chicks.

Most of our hens didn't hide their eggs. We spotted them right away. But Chicken Little was from a different group. She wasn't a domestic hen, born and raised in human conditions. She was among those we called The Jersey Girls—hens who had lived freely in wild vegetation in central New Jersey. The Jersey Girls were very serious about having children. They hid their eggs where they were difficult to find.

Chicken Little's offspring were a male and a female. Our staff named them Chuck and Charlotte.

Chuck was a healthy, handsome boy. Charlotte, however, was born with a deformed leg and was exceptionally small. Chicken Little was especially protective toward her. Whereas Chicken Little usually allowed young Chuck to venture a good distance away from her, she always kept young Charlotte nearby.

Whenever she felt Charlotte went too far, she went over to her, gave her peck, and nudged her back.

Charlotte never became as large as other hens, but she was a spunky adult. She energetically hopped about, exploring the pastures with the other chickens. And when other chickens came near her food, she pushed them away. All the humans on our farm admired and adored her.

Adult roosters are protective of hens, and Chuck was no exception. As an adult, he always kept an eye on the whereabouts of Chicken Little and Charlotte. When, after two years, Chicken Little died, Chuck watched over three other hens as well. With the chivalry of a true rooster, Chuck made sure the hens ate before he did. He and Charlotte also took a friendly interest in Violet, an elderly dwarf Nigerian goat who slept near them.

At the age of five years, Charlotte began to weaken and spend more time resting in the barn. Then one morning, one of our staff members discovered that Charlotte had died. The goat, Violet, was curled around Charlotte's body.

Violet's position was unusual. All our staff believed that Violet moved next to Charlotte during the night to provide comfort and protection.

Our staff wrapped Charlotte's body in a blanket and took her into the cottage until a grave could be dug. But when one member of our staff, Donna, returned to the barn, she saw that Chuck had entered it and seemed disoriented. He stood staring at a wall.

Donna wondered if Chuck might need time to mentally process the fact that Charlotte had died. Donna therefore brought Charlotte's body back into the barn. Chuck walked over and looked at it for fifteen minutes. Then he crowed twice, walked away, and resumed his normal behavior.

Chuck's response to Charlotte's death is akin to that of crows, elephants, and other animals. They, too, spend time watching over the recently deceased. They seem to need to engage in the kinds of behavior that humans call wakes and vigils. Just like us.

Chicken Little and her chicks

14

REGGIE

A Fierce Rooster

Our roosters' temperaments have varied. One, Silkie, was friendly toward everyone. He stood completely still while children and adults rubbed the feathers on his back. Our other roosters have had, to differing degrees, an aggressive streak.

The ferocity of one rooster, Reggie, is a legend on our farm. Abandoned in a park in Yonkers, he spent two months in a rescuer's basement before he came to us. On our farm, Reggie battled every rooster who came near him, and he went after us as well. He plowed into our legs like a football player making a tackle.

Sometimes, Reggie hid behind the barn and ambushed us when we entered his area. Everyone had to be on the alert and ready to step out of his way. One staff member, a college student named Chris, carried a garbage can lid with him to use as a shield.

But Reggie also exhibited a different, chivalrous side that is common to roosters. When Reggie saw food, he called the hens to it and never ate until all the hens had finished. When his group ventured into an outdoor pasture, he stood guard while the hens joyfully foraged in the grass or fallen leaves. Then, when he felt it was time, he directed the hens back to the safety of their aviary.

One afternoon a turkey, a tough female, wandered near him. Just as the two were about to fight, I grabbed the turkey and carried her away. She let me carry her because we had known each other since she was a baby and she trusted me. After that, Reggie never attacked me again. I can't be certain why his behavior changed, but he seemed to have gained a new respect for me.

One day, after three years with us, Reggie suddenly keeled over and died. His death was completely unexpected. Those who saw him an hour before said he was as vigorous as ever. The vet told us he was pretty sure it was a heart attack.

Equally surprising was everyone's response to his death. We were all sad. He had given us nothing but trouble, but we missed him. Chris, who was assigned to much of Reggie's care and had to deal with much of his aggression, still loves to talk about him.

Reggie

15

LEO

A Pot-bellied Pig

The first pig we adopted was a male of the pot-bellied variety. These pigs are related to the standard breeds but are smaller—usually growing to about one hundred pounds, whereas many regular pigs weigh over five hundred pounds. Our pig was particularly small because he was only two months old.

He was black and white and very cute. All our staff members adored him, and he liked everyone. When people gave him a pat on the back, he rolled over for a belly rub. The word soon spread in the community that there was a precious little pig at our farm, and visitors poured in.

The little fellow was brought to us by a family in Queens who had purchased him as a pet. The family and their pig became very attached to each other. But the family had to leave him alone during the day, and he cried. The noise disturbed the neighbors, who complained to the landlord. The landlord found out that Queens, like the rest of New York City, forbids residents to keep pigs, and he gave the family two weeks to remove him. Otherwise, he would evict them.

The family felt terrible. They didn't want to send the pig to a shelter, where he would probably be euthanized, but they couldn't find him a new home. As the landlord's deadline approached, the family thought they had found someone to take care of him, but the prospect fell through. The family contacted us and told us their plight. Our farm wasn't set up to care for a pig, but the situation was so urgent that we said we'd adopt him.

The family put the pig in their car and drove him to our farm. The mother, however, stayed home. She felt the trip would be too emotional for her. She was certain she would fall apart.

Because the pig was black and white, the children had named him Oreo, but the parents had told the kids that the name was temporary. The father told us we should feel free to select another.

Our staff disliked the idea of naming an animal after a cookie. After a discussion, everyone agreed to name him Leo after Leo Tolstoy, the Russian writer, pacifist, and vegetarian.

During his first month with us, Leo spent most daytime hours energetically exploring his new surroundings. He didn't show obvious signs of missing his family. But bedtime was another matter. When we put him into his stall for the night, he cried and cried.

Ellen went into his stall every evening to comfort him. She sang to him and rocked him to sleep. During the first few nights, she reached over and placed him on her lap. Then Leo started going over to her and climbing in her lap on his own volition. Finally, after an entire month, Leo crawled into the hay and went to sleep by himself.

A month later, Leo's former family came for a visit. This time, the mother came as well. It's difficult to know if Leo recognized them. He looked at them and then continued grazing in the grass.

I expected the mother to approach Leo, pet him, and say things like, "Do you remember me?" Instead, she sat on the ground about 10 feet from him and watched him quietly. She remained sitting for about 45 minutes, barely moving. She simply beamed.

As the family was getting ready to leave, I asked her what she thought about her visit. She put her hand on her chest and replied, "It was wonderful. It felt so good to watch him and see that he is so happy."

After their visit, I thought about why the mother sat at a distance from Leo. My guess is that her behavior was part of a difficult emotional transition. She had to let go of some of her close emotional attachment to Leo before she could even visit him. Then, by sitting apart from him at the farm, she could maintain some feeling of separation—while still loving him.

Leo

16

CHAVA

A Hen Who Wouldn't Give Up

A rabbi was walking along a busy street in Brooklyn when he saw a small hen being carried to a nearby slaughterhouse. He purchased the hen and announced to everyone within earshot, "I am saving this hen's life because all life is sacred."

A young woman who heard the rabbi was deeply moved and wanted to help. She asked him if she could take the hen to her apartment and look for a permanent home. He agreed. To honor his pronouncement about the sacredness of life, she named the hen Chava, a word for *life* in Hebrew (English speakers pronounce the word "hawvuh"). When the woman called us, we adopted the hen. For the next two years, Chava lived happily with the other chickens. She enjoyed foraging in the grass and taking dust baths in the dirt. On nice days, she liked to sit and bask in the sunshine.

Then one day a staff member noticed that Chava was unusually quiet. He told this to our chief caregiver, Joy, who took the hen into the cottage and placed her on our examining table, where Joy detected a foul odor. She lifted Chava's wings and saw a huge cut on her abdomen. Some flesh had died, causing the odor. No one knew how Chava received the wound. All the staff wished they had noticed it earlier.

I rushed Chava to the vet. He told me she was so badly injured that she probably wouldn't survive. He suggested I consider euthanasia. I said our sanctuary was filled with animals who had somehow escaped death, and I wanted to give her a chance. I took her back to our sanctuary.

Adhering to the vet's instructions, we kept Chava in a large crate in our cottage. We soaked her wound, applied antibiotic cream, and changed her dressing every day. She was in surprisingly good spirits.

We took her to the vet for a two-week follow-up visit, where he carried her into the procedure room. When he returned to meet us, he looked sad. "The dead tissue is stuck to the wound," he said. "The wound cannot heal." We were heartbroken.

Suddenly, his assistant appeared and said, "Doctor, would you please come back to the procedure room with me?" When they returned, they were beaming. The assistant had noticed a tiny spot where the dead tissue might be peeling off. The vet then discovered that a good amount of it was loosening. Chava was healing!

The vet sent us home to continue Chava's treatments. After four months she completely recovered.

Over the years, Chava had other setbacks. She acquired a sinus problem which, despite months of treatment, never completely went away. But it didn't slow her down. She continued to actively forage for food and run in the grass.

In her last year, Chava suffered from severe arthritis. It was so bad in one leg that she couldn't put weight on it. But she kept moving. She hurled herself forward and hopped on her good leg, sometimes flapping a wing for extra speed. Chava exemplified a life force that wouldn't be denied.

Chava

17

CESAR

A Mischievous Goat

Late one night, Ellen and I were awakened by a phone call from a police officer in the Bronx. He asked if we would adopt a small male goat being held in his precinct.

The officer explained that he and his colleagues had been sitting around the precinct that night, mourning the fatal shooting of a fellow officer, when a resident brought in the goat. The goat had been roaming the streets.

The officer said he loved animals and worried that if the goat went to a city shelter, he might be euthanized. The officer added, "After my colleague's death, I can't stand the thought of more dying? Will you take him?" We said we would, and he and his partner drove the goat to us early in the morning.

When the goat got out of their car, we saw he was just a baby; his head only came up to my knee. He was thin and frightened, and his head and horns were covered with wax, suggesting to the officers that he had escaped a ritual slaughter. Our vet immediately came over and told us the goat was anemic and had pneumonia.

But despite his fears and ailments, the little goat began to show signs of a lively curiosity. We named him Cesar, in honor of a cheerful construction worker who has made numerous repairs on our farm.

Cesar soon recovered from his illnesses and became one of the most adventurous and fun-loving animals on our farm. He goes everywhere. He leaps over stall gates, climbs stairs, and jumps on any structure that will take him to new places. He also likes to cause problems.

One morning, one of our staff members came over to Ellen and asked, "Why is the ceiling fan in the barn running?" Ellen didn't know. After all, it was a bitterly cold winter day—much too cold for the fan to be on. She turned it off.

Later that day, Ellen heard the fan running again. Puzzled, she entered the barn. She saw Cesar, high up on a narrow ledge, pulling the fan's cord with his teeth. He was even varying the fan's speeds. She found the sight amusing, but it took considerable effort to get him down.

Cesar likes to look in our pockets, which interferes with our chores. It's difficult to carry a bucket of water with a goat pressing against a pant leg.

When visitors come, Cesar hangs around the gates and sometimes pushes his way through. He will run into a chute and resist all efforts to steer him back to the pasture. Only after we give up and leave him in the chute will he decide to return.

A local television reporter came to interview me about our farm. She thought the barn would make a nice background, with chickens and goats milling nearby. She hooked me up to a portable microphone, and I began talking into the camera, when I felt an unusual sensation at the bottom of my shirt. I looked down and saw that Cesar had chewed up the microphone cord. That ended the interview.

I sometimes lose my composure and say things like, "Darn it, Cesar, you're being a real nuisance." But other staff members, especially young men like Chris, sympathize with him. One day Chris was overheard telling him, "Don't worry about it when they get mad at you, Cesar. You're a good boy. You just can't help getting into trouble sometimes."

Cesar

18

BOGIE

A Rooster Makes a Great Escape

Bogie began life as a middle school hatching project. A teacher had ordered six fertilized eggs from a company to show students how chickens hatch and grow. The teacher believed there would be homes for the chicks once the school year was over, but as often happens, she couldn't find any.

One option was to return them to the company, but the company was likely to destroy them. Fortunately, a parent contacted our farm animal sanctuary, and we adopted the chicks.

Ellen and I usually let our staff members name new animals, and they decided to name this group of chicks after classic movie stars. They called this young rooster Bogie, after Humphrey Bogart.

When the staff named him Bogie, they didn't think of him as a tough guy like the characters played by his namesake. But as the rooster grew, he began living up to his name. He took a bold stand in disputes with other roosters and liked to fly to the top of a fence and crow, as if announcing his importance. Sometimes he flew over the fence to explore the surroundings, and we had to guide him back into his fenced area.

Then one day, before we could guide him back, he was suddenly gone. All we saw were piles of his white feathers on the ground and scratches in the dirt, signs of a struggle. Bogie had been snatched by a predator, probably a hawk.

With heavy hearts, we put the other animals to bed and went to our own houses for the night. Later that evening, as rain came down, I performed the routine bed check to make sure all the animals were secure inside their enclosures. As I looked around, I half expected to see Bogie, but Bogie was gone.

The next morning, Ellen went to the barn to feed the chickens breakfast and let them out for the day. Then she came running back to our house and exclaimed, "You'll never guess what happened! Bogey is back. He was standing by the barn when I went to feed the chickens. He must have escaped the hawk and hid during the night."

I hurried to the barn and there he was. He was wet and missing several feathers, but alive and well. We were all overjoyed.

Bogie's disappearance showed us that we had to take further precautions against predators, and we added netting and wire to several open areas. Bogie also demonstrated that he was, and is, tougher than we had ever imagined.

Bogie

19

MR. T

A Tamworth Pig

While driving, I frequently pass a small pen for holding farm animals. Usually, the pen is empty, but one day I saw it holding a large pig. I stopped and asked the farmer about the pig. He told me the pig, a male Tamworth, had been the sole remaining animal at a breeding facility that recently shut down. He said he was keeping him a while before he went to slaughter.

I told Ellen about the situation. At the time, our farm sanctuary didn't have room for a large pig, but we wanted to save his life. We asked the farmer if he would give him to us if we could find him a home. The farmer was happy to oblige, but he said we needed to find a home within two weeks.

We called the Tamworth pig "Mr. T." We frequently visited Mr. T in his pen, and he was very friendly. Whenever he saw us coming, he hurried over to the fence for petting.

But numerous phone calls to potential homes came up empty. Two weeks went by, and we had to get a week's extension to continue our search.

More phone calls were also unsuccessful. We became very worried. Would we ever find Mr. T a home? Could we save his life?

Finally, we got some good news. A pig sanctuary north of Syracuse, The Tusk and Bristle Sanctuary, said it would foster Mr.

T. The sanctuary was in the process of closing, but the owner, Carol, said, "Don't worry. We'll keep Mr. T alive until you find a permanent home. Bring him up to us." Carol's response brought us enormous relief.

Our neighbor owns a horse transport firm, Morrissey's Horse Pullmans, which drove Mr. T to The Tusk and Bristle Sanctuary. There, Mr. T was housed with a young female pig named Matilda. Matilda had come from a farmer who didn't want her because she was, in his words, "too clingy." She pestered him for so much attention that he couldn't do his chores. At Tusk and Bristle, Matilda and Mr. T became close companions.

But the pressure to find Mr. T a permanent home remained. And now there was a complication. We didn't want to separate Mr. T and Matilda, so we looked for a home that would take them both. Once again, phone calls produced no results.

Then Ellen noticed that Richard Peppin, a family friend and animal lover, had sent an email with cute photos of piglets at the bottom. Ellen wondered if he liked pigs so much that he'd want to help us. She called him, and he suggested we call the Poplar Spring Animal Sanctuary in Maryland. The sanctuary responded positively and is now the home of two very happy pigs.

You have heard the saying, "It takes a village to raise a child." The rescue of Mr. T and his friend Matilda required help from so many people that we might say, "It takes a village to rescue two pigs."

For more on Mr. T and Matilda at Poplar Spring, see this article in *The Dodo* newsletter: https://www.thedodo.com/on-the-farm/clingy-rescue-pig-bff-boar.

Mr. T and his small friend, Matilda

20

MILO

A Goat Defends His Mother

Two of the goats living on our farm are Bessie and Milo, a mother and son. They came to us when Milo was a baby and was still nursing.

The two goats had been in a backyard in Coney Island when a neighbor learned that their owners planned to have them slaughtered for a barbecue. The neighbor became distraught and called the police. The police informed the owners that it is illegal to keep farm animals in Coney Island and took the goats to an animal control facility. They remained there for several days while the staff searched for a permanent home. We welcomed the goats and were immediately impressed by their sweet dispositions. They were very friendly toward humans and enjoyed our pats and rubs. One of our youngest staff members enthusiastically suggested that we call them two of his favorite names, Bessie and Milo.

Milo grew to be much larger than Bessie, but like her, he has remained a gentle soul. Both will sometimes play fight with our other goats, butting heads with them, but never very roughly.

And Bessie and Milo, like all our other goats, try to avoid Duncan.

Duncan is the dominant goat on our farm. He is very strong, and although he has calmed down a bit over the years, he can still be a bully. He sometimes tries to start aggressive play fights with other goats just to demonstrate his superior strength.

One summer day, Duncan started a tussle with Bessie. Bessie tried to retreat, but Duncan kept after her. She looked very upset. I was thinking about intervening when, from afar, Milo raced over and picked up the fight with Duncan. I was amazed to see Milo butt heads so forcefully. Over and over, Milo and Duncan rose on their hind legs and hammered down on each other with their horns. When their horns hit, the cracking sound rang throughout the farm.

After several minutes, both goats decided the fight was over. The contest ended in a draw, and Duncan and Milo went their separate ways.

Four years have now gone by, and Milo and Duncan haven't battled again. But Duncan hasn't pestered Milo's mother again, either.

Milo and Bessie

21

A RED ROOSTER TURNS BLUE

A small young rooster was brought to us by an off-duty policewoman. She had spotted him hiding near a live meat market in Queens. The chicken had somehow escaped the building and his slaughter.

The policewoman told Ellen that she rescued the chicken because she loves animals and named him Big Red because of his bright red coloring. Ellen asked why she called this chicken Big Red. The woman just shrugged and said she liked the name.

She wasn't sure that the chicken was a rooster or hen, and we couldn't tell, either. It is difficult to determine chickens' sex when they are young.

On our farm, Big Red grew and grew; he eventually weighed thirteen pounds, heavier than any chicken I have seen. He certainly lived up to his name. And he soon began to crow and tried to mate with the hens. He was a rooster for sure.

One afternoon, after Big Red had been with us three years, a crisis occurred. Chris and another young staff member, Ryan, came running over to Ellen and me. Chris was carrying Big Red. "Look, Big Red is blue," he shouted. "He's gasping for air."

"He cannot stand up," Ryan added. "I think he is dying."

The rooster's crown and the red skin beneath his beak had indeed turned blue. I drove him to an emergency veterinary clinic, where he was put in an oxygen room overnight. In the morning, the vet phoned to say the oxygen had helped a lot. He had regained some of his red coloring and was even walking about. I drove to pick him up, full of hope.

But when I arrived at the clinic, the vet told me Big Red had taken a turn for the worse. When I saw him, he was sitting and looked weak. His coloring didn't look good, either. The vet asked me what I wanted to do—meaning did I want her to euthanize him. She relayed a message from Ellen: Because I was at the clinic and could observe Big Red, the decision was mine.

For some reason, I felt I shouldn't have him euthanized. I drove him back to the farm and put him in his aviary.

And to our amazement, Big Red stood up and began walking about. He was redder, too. Within an hour, he was crowing. He was back to his old self.

The vet didn't know the cause of his sudden oxygen loss or how he recovered. Ellen, a pediatrician who has acquired some medical knowledge of farm animals, speculates that part of his heart went into a spasm, blocking the circulation of blood and oxygen. The extra oxygen in the clinic helped him recover, as did his ability to relax back home on the farm.

Big Red

22

BRIGITTE & HARRIET

Inseparable Hens

A man called to see if we would adopt two young chickens. He said his mother-in-law bought them from a mail-order company. She had grown up with chickens in the Philippines and thought his new family would enjoy them, too. But the family couldn't find time to care for them, so the mother-in-law was about to return them to the company. Fearing that the company would simply kill them, the man reached out to us and was relieved when we agreed to adopt them.

One chicken was an ordinary-looking brown hen who was exceptionally bold. No other chicken, not even a rooster, could chase her away. She always stood her ground. She sometimes even pecked the larger male. In honor of her bravery, we named her Harriet after Harriet Tubman, who took great risks helping slaves escape to the North through the Underground Railroad.

The other chicken, a black-and-white hen, was strikingly beautiful. We named her Brigitte, after Brigitte Bardot, the famous 1950s and '60s French movie star who later became an animal rights activist.

Harriet and Brigitte ventured far and wide. They jumped up on fences, roamed across pastures, and slept on the high rafters in a barn. And they always went together. They were inseparable.

Harriet and Brigitte also became our companions during our chores. Late every afternoon, after we put the sheep and goats into their stalls for the night, we rake up the poop in the pastures. Harriet and Brigitte consistently accompanied us. As soon as they saw us head out to a pasture, they eagerly ran over to join us.

While we raked, they foraged for plants and insects, scratching the dirt, grass, and leaves. From all appearances, they seemed so engrossed in their foraging that they had forgotten all about us. But whenever we moved to a new pasture, they promptly followed.

The two chickens' enthusiasm for life was a joy to behold. But after five years, Harriet suddenly became weak and died. Our veterinarian found no specific cause and suggested that she simply died a bit younger than we expected.

All the humans at our farm were distraught. Brigitte didn't react in any dramatic way, but she moved around a bit more slowly and her trips to the pastures became infrequent.

Six months later, three more chickens died within a two-week span. The deaths occurred so close together that we worried about a contagious infection. True, the three chickens were quite old, so it was likely that they died of natural causes, but until necropsies confirmed the absence of an infection, we were shaken.

During this stressful time, Brigitte resumed going into the pastures with us. She accompanied us every day. I doubt we will ever know her motive for joining us, but her constant presence helped us. We felt she was saying, "I'm with you. Carry on."

Harriet and Brigitte

23

Ethel and Our Cows

In November, 2019, we moved to a larger farm a mile down the road. With cars, trucks, and vans, we transported all the animals to their new home in one morning. The move was stressful for some of them, but everyone soon enjoyed their spacious new surroundings.

We were especially excited by the new farm because its expansive pastures would enable us to adopt both cows and large pigs—animals who need lots of land. The new property already had two barns, and we built two more: one for the cows and the other for the pigs.

A year after the move, we adopted two cows from a small dairy farm in Connecticut. A couple had run it for decades, but the husband had recently died and his wife's physical problems prevented her from taking over.

She had sold all her cows except these two: a female named Ethel and her daughter Surprise. Ethel was nineteen years old, which was about her expected lifespan. Surprise was thirteen. The woman and her husband had developed a special affection for them, and she wanted them to continue to live together. But other dairy farms didn't want a cow Ethel's age.

When Ethel arrived, she had a growth on her eyelid. Our vet removed it, but he discovered a year later that she had cancer beneath the eye. He told us it couldn't be treated, and he doubted she would live very long. But over a year has gone by, and although old age has slowed Ethel down, she is still active.

We adopted six more cows, four males and two females, including four-year-old Flo. The farmer didn't want her because it seemed that she couldn't get pregnant, and cows must have babies to make milk.

It struck me that Ethel, the oldest cow, was particularly attentive to the new arrivals. When, for example, Flo arrived and displayed confusion, Ethel seemed to guide her to the barn. I thought Ethel might be acting like a herd's matriarch, taking responsibility for the newcomer, but I wasn't sure.

Once we had adopted these cows—eight in all—we decided we had reached our limit. We couldn't care for more. Then our vet told us another was on the way: Flo was pregnant. Despite what we had been told, she had conceived at her prior residence. Not long after our vet's examination, Flo gave birth to a healthy baby boy, whom the staff named Cole.

Watching Flo care for her baby was a special sight—and one that is rarely seen on a dairy farm. The farmers don't want the cows to give milk to their babies—they want to sell it to humans—so they separate the mothers and babies within a day after birth.

Flo nursed Cole and treated him with great affection, frequently licking him and snuggling with him.

Cole enjoyed being with his mother, and he also liked to play. When he was only seven days old, he ran back and forth beside the barn and tried to play with a tree, dancing around it and butting it. Because there were no calves his age, we tried to play with him when we could.

But one night, when Cole was four months old, his playfulness gave me a real scare.

One of my bedtime chores is to climb up a hill with a flashlight and make sure the cows' water trough is full. That night, as I was walking back down, Cole came running up toward me, eager to play. At this time, he weighed about 400 pounds, and I was caught off guard. I couldn't find solid footing on the rough ground of the steep slope. I knew Cole only intended to play, but his head-butts could easily knock me down. "No, Cole!" I shouted. "No Cole, I can't play!" "No!" But he persisted.

Then I heard heavy footsteps behind me. Someone very large was gradually approaching. It was Ethel! She looked at us for a minute, paying most attention to Cole. Then she lumbered back down the hill.

Once again, I had the impression that Ethel acted as the responsible matriarch. My shouts prompted her to see if Cole was in danger, even though the trip up the hill wasn't easy for her. Ethel assumed it was her duty to assess the situation, and she concluded that everything was all right.

Cole still wanted to play once Ethel left, but I made it safely down the hill. Since that night, I have always made sure to have a good path down the hill.

Ethel

24

A Pig Rescue

Late one fall afternoon, we received phone calls about two pigs wandering beside a heavily trafficked road about ten miles away from us. The callers were worried that the pigs would be hit by a vehicle.

I drove out to where callers had spotted them, but the pigs were nowhere in sight. After an hour of searching, darkness fell, and I decided that there was no point in looking any more. Dejected, I walked back to my car.

Then, just as I was about to get in my car and drive home, two young adults approached and began to unlock the car next to mine. On a long shot, I asked, "You haven't by any chance seen two pigs, have you?" They said, "Yeah! We saw them on our hike! They were a half mile down the railroad tracks. They were next to the tracks, half covered in leaves. It looked like they were bedded down for the night."

Early the next morning, Joy (our head caretaker) and I drove back to the area where the hikers had seen the pigs. When we arrived, daylight was just breaking. We saw the pigs and waited until they woke up.

Both were females, with unusual hair. It curled and looked like wool. We later learned that they are a breed that is informally called a "sheep pig." The breed's technical name is *mangalica*, and it originated in Hungary. Sheep pigs are like other pigs in every way, except for their woolly hair.

The two pigs were friendly to us, and they eagerly drank the water and ate the food we brought them.

After eating, they walked up a hill into the woods. Then, after just a few minutes, they returned and ate some more. I couldn't think of a reason for their excursion. But when I retraced their route, I saw their stools. Even out in a strange new place, the pigs didn't want to poop in the same place they ate.

We wanted to take them to our sanctuary, where they would be safe, and we tried to lure them into my Honda Element with a bit more food. But the ramp was too narrow for them to comfortably climb. They walked half the way up, then turned back.

Joy and I thought about our neighbors, who own Morrissey Pullman horse transport. Their trucks would have wide ramps. Joy called them, and they were able to get a truck to us.

With apples as encouragement, the pigs went right into the truck and on to our sanctuary. The staff, thinking of movie characters who had been on the road, named the pigs Thelma and Louise.

We didn't know how the pigs got to the busy road. Maybe they ran loose from a farm, we thought, and the farmer would be looking for them. But despite phone calls to farmers and announcements on social media, no one claimed the pigs.

We speculated about other possibilities. Maybe they escaped a slaughterhouse. Maybe they were just abandoned. We had no way of knowing. But we were certain of one thing: They would be happy at our sanctuary, where they would live out their natural lives.

Thelma and Louise

25

HOUSTON

A Quarter Horse

A couple decided to purchase a farmhouse in rural New Jersey. On the property were two dogs, three sheep, and a horse named Houston. Before the sale was completed, the couple visited the property several times and saw that the dogs were often aggressive toward the sheep. When the sheep were frightened, they ran to Houston for protection, with one or two squeezing under him.

The couple learned that Houston was a Quarter Horse, a breed that cowboys often ride in the West. He was elderly—the seller estimated that he was thirty years old. This is the typical lifespan of this breed, although they sometimes survive another five years. He was skinny, with a white, off-center marking on his face.

Houston also was blind in one eye. The seller didn't know the source of this impairment. Although the horse could see well enough to get around on his own, he sometimes followed the sheep, seeming to use them as guides.

As the sale neared closing, the seller said he didn't want to keep his three sheep and horse. He said if the couple didn't purchase the sheep, he would send them to auction. His price for the three was $400. The couple, who love animals, knew that auctions often result in animals' deaths, so they purchased the sheep.

The seller had such a low opinion of Houston that he didn't even want money for him. He said Houston was so unruly "he can't be trailered" (meaning, led into a trailer for transport) and "he hates all men." He offered to give Houston to them for a dollar, simply to make the transaction official. Otherwise, he would have Houston euthanized. In fact, he had already dug a pit for the horse's burial. The couple paid the dollar.

The couple found a farm sanctuary for the sheep. After considerable searching, they also found a new home for Houston (and a transporter who succeeded in taking him to it). They thought their ordeal had a happy ending.

But the new owner wanted to use Houston for riding lessons, and he bucked off the first rider. Less than a year after he was adopted, the new owner called a vet, who scheduled Houston for euthanasia in one week.

The horse transporter, a woman named Lorraine, became actively involved. She made numerous phone calls and posted many messages. When news of Houston's situation reached us, we agreed to adopt him, and she brought him to us.

When Lorraine led Houston off her truck, he was nervous, but he looked fit for his age. Lorraine handed him off to a staff member, who walked him around a bit. Then I asked the staff member to let him run loose. Holding his head and tail high, Houston trotted in wide circles with speed and grace. His stride was beautiful.

Houston was very wary of us. A staff member, who had considerable experience with horses, was able to get his halter on and off, but only through deception. She gave him pieces of candy while she hid the halter, then threw it over his head when his attention was diverted.

We needed to get a halter on him to groom him. A halter was also necessary for any veterinarian examinations and foot care. But we felt it was important that he come to trust us, and we didn't want to use tricks.

Ellen therefore tried a different approach. She first let Houston get to know her, just standing around him. Whenever he came over to her, she gave him a piece of carrot. After several days, she tried touching his neck with the back of her hand. Whenever he resisted, she immediately backed off. If he did let her touch him, she gave him a bit of carrot. After three months, with this gradual approach, he let her pet him and put a halter on him.

Houston quickly made friends with our two donkeys, but he was nervous around some of the farm animals. This was especially the case with the pigs, even though they were separated from him by a fence. When the pigs first came within a few yards of him, he became excited and trotted rapidly in large circles, just as he had on the first day with us.

Houston's stride was magnificent, like that of an Arabian stallion. Watching him run, it struck me that he was saying, "I'm fast and powerful, so don't mess with me!" And I thought how wrong it had been for people not to respect him. Houston was a wonder to behold.

Houston

26

EMMA

Our Sweetheart

Emma, one of our four original turkeys, was exceptionally nice to humans. She frequently approached us and sat quietly while we patted her.

Occasionally, another turkey became angry at a human and pecked the person. Whenever Emma saw this, she pecked the turkey, seeming to say, "Stop that! Be nice!"

As the years passed, Emma's three original friends died, all at about six years of age. Six years is a rather long time for a domestic turkey to live, but Emma kept going. New turkeys came, and Emma was friendly to them all.

Finally, at the age of nine, she developed serious health problems. Her eyesight weakened, and she experienced respiratory difficulties. She also developed arthritis in her legs, which restricted her mobility. All the humans on our farm were so devoted to Emma that they consistently looked for ways to improve her health and comfort. They looked for medications to ease her pain and gave her a separate place to eat.

Our oldest goat, Basil, demonstrated her own concern for Emma. One day, Emma was eating out of her bowl when Gracie, a pot-bellied pig who had recently joined our farm, rambled over and began to eat some of Emma's food. Seeing this, Basil rushed to the scene and butted Gracie away. Emma then finished her meal in peace.

Emma's respiratory problems gradually worsened, until suddenly, she could barely breathe. In desperation, Ellen and I drove her to the vet, but he couldn't save her. She was eleven years old, an amazing age for a domestic turkey.

We drove her body back to the farm, buried her in our backyard, and gathered the staff for a tearful ceremony. That evening, I went into Emma's barn as part of my routine bed check. When I walked inside, the turkeys were squawking loudly. Without thinking, I said in a sad tone, "Emma died." They suddenly fell silent, and they were still quiet when I checked an hour later.

What had transpired? This might sound farfetched, but here is my guess: The turkeys were squawking because they were upset about Emma's absence. They wanted to know what happened to her. When I spoke, they understood from my tone of voice that Emma had passed away, and they became sad and subdued.

The next morning, Joy, our head caretaker, also had an unusual experience. When Joy went to let the goats out of their stalls, Boomer, the goat who slept nearest to Emma, didn't get up. When Joy knelt beside him to see what might be wrong, he nuzzled his head against her. She patted him for a couple minutes, and he then rose to his feet and went outdoors. Joy felt that Boomer just wanted to be comforted.

Other staff members independently told me that the day after Emma's death was unusual. All the animals—the turkeys, goats, chickens, sheep, pigs, and others—were exceptionally quiet and gentle. After losing Emma, no one was in a mood to quarrel.

A child hugs Emma

27

In Praise of Roosters

People are increasingly raising backyard chickens. They enjoy the animals and their eggs. Caring for them is a gratifying, home-based activity.

But municipal governments don't welcome all chickens. Many allow hens but not roosters because they provoke complaints about crowing.

Bans on roosters, which are prevalent across the country, have created a huge problem. Our farm animal sanctuary hears about it several times a week. People call to explain that they ordered only female chicks from a company, but one or two grew into roosters. (This outcome is common because hatcheries have difficulty distinguishing between the males and females when they are young.) The callers say they would be happy to keep the roosters, but their local governments insist they get rid of them. They don't want to simply dump the birds in a park or along a roadside, and they fear that if they return them to the seller, they will be killed. They hope we will adopt them.

But our farm sanctuary, like others we know, has taken in all the roosters we can care for. We have recently built additional space for them, but the problem is far too vast for sanctuaries to handle.

The municipal bans have added to the grim fate of roosters in general. Those born into the U.S. meat industry typically spend their lives crowded together in huge windowless sheds, just like the hens. Baby roosters in the egg industry aren't permitted to live at all. Because they don't lay eggs, they are killed within a day of hatching.

I believe that the more that people learn about animals, the more they will appreciate them and want them to have full and happy lives. With time, they will even develop positive attitudes toward roosters and their crowing.

Roosters are tough fellows. A few on our farm have fought each other so fiercely that they have drawn blood. We had to build separate aviaries for them, placing each with a separate group of hens. Over the years, three of our roosters even attacked us and our staff members. They plowed into us like football players making tackles.

Although the bird's fighting temperament causes problems, it also serves to protect the flock. I witnessed a stirring example of this.

While driving my car, I saw a large pen with chickens inside. To my surprise, there was a raccoon inside as well. The hens were all huddled against the back fence. Then the rooster stepped up to the raccoon and the two stood face-to-face. It was as if the rooster was saying, "If you think you're going to get those hens, you have to go through me!" A rooster is no match for a raccoon, but this one's bravery was something to behold.

This face-off continued long enough for me to get out of my car and throw a pebble toward the raccoon, which scared him away. I then informed the owner so he could buttress the pen against predators.

Neighboring farmers have told me of similar incidents. As a fox, raccoon, or other predator approached the hens, the rooster intervened. He lost the battle and died, but while the fight took place, the hens had time to escape.

Like brave medieval knights, roosters are chivalrous. When they spot something good to eat, they call the hens to it, and they don't partake until the hens are finished. When their group ventures into an outdoor pasture, they stand guard while the hens eagerly forage. Then, when he feels it is time, he directs the hens back to the safety of the aviary.

The rooster's crow, which is the biggest problem for many people, is part of the animal's bold nature. It is never half-hearted. The animal rises, flaps his wings, and calls out with all his might. Children who visit our farm are thrilled by it. Many try to imitate it.

Our roosters crow at dawn and throughout the day. Before Ellen and I started our farm, we expected that the sound would disturb us. And our roosters did wake us up earlier than we wished. But this only happened the first two mornings. Moreover, we soon began to feel that their crowing is somehow uplifting, and this feeling is shared by those who have worked with us. The rooster seems to be proclaiming his toughness, but he also conveys something more fundamental. Henry David Thoreau said the bird's high-spirited strain expresses the "effervescence of life."

I suspect that people who favor town bans on roosters have endured many noxious mechanical sounds, like lawn mowers and leaf blowers, and they don't want to be disturbed by roosters as well. But the rooster's crow isn't part of the mechanical world. It comes from nature. The rooster is nature's trumpeter, sounding out nature's force and vitality. He calls attention to the miraculous world of living things. I sincerely hope people who have supported bans on roosters will reconsider.

A rooster waits for hens to eat

28

THE IMPORTANCE OF MOTHERS

Every spring, our farm sanctuary receives one or two calls a day from people who would like us to adopt their young chickens or ducks. They usually purchased the animals from a farm supply store like Tractor Supply. The babies looked so cute under heat lamps that the customers couldn't resist taking them home. But they discovered that caring for the animals was more trouble than they anticipated.

In June, we also receive several calls—about twice a week—from schoolteachers. They explain that they ordered fertilized eggs from a company to show their children about hatching and development. The children watched over the eggs and babies, keeping them warm under heat lamps. But now the school year is ending, and the teachers cannot find permanent homes for the hatchlings. They, like other callers, hope we can take them in.

But our sanctuary is typically filled to the brim with chickens and ducks. We rarely have room for more. Other sanctuaries are usually at full capacity, too. We give people ideas for finding homes, but many chicks and ducklings are simply abandoned. People tell us that they have seen them on the roadsides and in the woods.

Our sanctuary, working with Dr. Karen Davis of United Poultry Concerns, is trying to make the public aware of the fate of so many of these young animals.

Inspired by a concern raised by Dr. Davis, our sanctuary also encourages people to question the widespread human practice of raising babies without mothers. On our farm, a few of our ducklings and chicks have had mothers, and these babies have shown us what the others have missed.

For one thing, mothers provide warmth and comfort. Heat lamps can keep babies warm, but they cannot provide the full comfort a baby gets when nestled under a mother's feathers. True, motherless chicks and ducklings huddle together, but they are too small to substitute for a mother's body.

In addition, mothers offer protection. When a baby dives underneath a mother's feathers, the baby hides from threats in the surroundings. A baby under a heat lamp is always exposed.

As babies grow, mothers protect them by making sure they don't roam too far. You might recall (in Chapter 13) how Chicken Little kept her tiny and lame chick, Charlotte, very close by. When she judged that Charlotte had wandered too far, she rushed over, gave Charlotte a peck, and nudged her back.

In the barn, the mother's actions kept Charlotte from venturing into areas where a turkey or goat might inadvertently step on her. Outdoors, Chicken Little kept Charlotte from going so far that she couldn't get back if a predator appeared.

Animal researchers have found that chicken hens display an innate rescuing instinct, and we also have seen this on our farm. When a chick has fallen or got stuck in a bush, the chick has emitted a high-pitch distress call that prompted the hen to rush to the baby's aid. Our motherless chicks are more helpless. They are more alone in the world.

Teachers want students to learn about animal birth and growth. This is a great goal, but students can only learn so much from motherless babies. They don't learn how babies hide in their mother's feathers, how mothers keep their babies nearby, or how mothers rescue them.

Although a farm sanctuary offers many opportunities to learn about animal behavior, the ideal setting is a species' natural environment—the wild. On our farm, the setting that is the most open and wild is our large pond. Among the animals there, the most visible are mallard ducks. Many were raised in artificial conditions and escaped a hunting club, but they often regain their natural behaviors on the pond. I have spent hours watching them.

On several occasions, I saw a mother lead her ducklings to a shore where they energetically pecked for food. They acted like they were starving. But sometimes, for reasons unknown to me, the mother suddenly entered the pond and began to swim away. Every time, the ducklings stopped foraging and followed her into the water. They obviously wanted to eat, but the urge to follow their mother was stronger. If we were able to hold a conversation with them, I'm sure they would tell us, "Nothing is more important than my mother."

Ducklings follow their mother

29

What I Have Learned

When we opened the sanctuary, I wanted to learn about nonhuman animals. I wanted to gain knowledge that would help me as a professor of psychology. And I soon found myself learning a good deal.

For example, many psychologists are interested in imprinting, which has to do with youngsters' attachment to a mother-figure. On our pond, I observed just how strong this attachment can be. The mallard ducklings' urge to follow the mother overrode their search for food. Whenever a mother left the shore for the water, her babies abandoned their foraging and went with her.

I also witnessed first-hand behavior that I had only read about, as when the ducks organized themselves into a battalion to chase off two large geese (a group action that animal researchers call "mobbing").

I soon learned, in addition, that animals have individual personalities and are full of surprises. For example, I was amazed to see the gentle goat Milo charge the dominant Duncan to protect his mother. And I was just as surprised by numerous other events, as when our goat Basil made sure that Emma, our elderly turkey, could eat without a pig's interruption.

Some behaviors were more than surprising; they went beyond what I thought possible. I never would have thought that our rambunctious young turkeys would quietly listen to Girl Scouts' solemn and reverent pledges. The turkeys seemed sensitive to humans' spiritual emotions—a sensitivity that I still find somewhat difficult to believe.

Even more extraordinary was the sight of our turkey Ducky walking out of her barn to meet a Buddhist monk. Before that, I had never seen Ducky approach any human—at least not in a friendly manner. Moreover, Ducky's arthritis made it nearly impossible for her to move at all. Why did she make such an effort to see the monk? The only explanation I could think of was that her act had to do with a past life—an explanation that still lies outside my Western scientific belief system.

I discovered, then, that animals present us with mystery. This insight is hardly new. Many people who closely observe pets and other animals have come across behavior that seems inexplicable. Scientists such as Albert Einstein and Rachel Carson have even maintained that nature's mystery will always be with us; for she works in ways that are beyond the full grasp of the human mind. And this mystery, these scholars have added, is of great value.

It keeps our sense of wonder and curiosity alive.

ACKNOWLEDGEMENTS

I wish to thank my wife, Ellen, and our sanctuary staff for their dedication and thoughtfulness. I also am happy to acknowledge Lantern Publications and Media for its fine work and commitment to the well-being of our animal relatives.

Many people have supported and encouraged our farm's efforts. I wish I could thank everyone, but the list would be too long. Here, I will mention Richard Peppin, Harold Hovel, Marc Bekoff, Karen Davis, Peter Fairweather, Joyce Friedman, Richard Lewis, Liz Goodenough, Shari Thompson, Abigail Canfield, the Sarah Lawrence College Child Development Institute, and my colleagues at CCNY.

The following media granted permission to reprint material in some of this book's chapters: *Dirt* magazine, part of chapter 26; All-Creatures.org, segments of chapters 27 and 28; Red Wheel/ Weiser, lines in chapter 2. Early versions of some stories appeared in the *Poughkeepsie Journal*.

The photo of Chicken Little was taken by Donna Scott, that of Mr. T and Matilda by Dar Veverka, and that of Houston by Teresa Auror. Adam Crain took the photo of Emma and a child, and I took the photos of the ducks. Ellen Crain took all the others.

BIBLIOGRAPHY

Carson, Rachel. *The Sense of Wonder*. New York: Harper Collins, 1998.

Einstein, Albert. "What I Believe." In *An Einstein Encyclopedia*, edited by Alice Calaprice, Daniel Kennefick, and Robert Schulmann, xxi-xxiii. Princeton, NJ: Princeton University Press, 2015.

Lorenz, K. *On Aggression*. Translated by Marjorie Kerr Wilson, 26-27. New York: Harcourt, Brace & World, 1966.

Rilke, Rainer Maria. "The Eighth Elegy." In *The Selected Poetry of Rainer Maria Rilke*, edited and translated by Stephen Mitchell, 193, 195. New York: Viking International, 1989.

Thoreau, Henry David. *The Heart of Thoreau's Journals*, edited by Odell Shepard, 199. New York: Dover, 1961.

Tinbergen, Elisabeth A. and Tinbergen, Niko. "Early Infantile Autism: An Ethological Approach." In *Advances in Ethology: Supplement to the Journal of Comparative Ethology, 10* (1972), 37.

ABOUT THE AUTHOR

 **BILL CRAIN** is Professor Emeritus of Psychology at The City College of New York. In 2008, he and his wife Ellen founded Safe Haven Farm Sanctuary in Poughquag, NY. In 2018, he received a PETA Hero to Animals award for his efforts to protect black bears from hunting. Bill and Ellen have three children and six grandchildren.

ABOUT THE PUBLISHER

LANTERN PUBLISHING & MEDIA was founded in 2020 to follow and expand on the legacy of Lantern Books—a publishing company started in 1999 on the principles of living with a greater depth and commitment to the preservation of the natural world. Like its predecessor, Lantern Publishing & Media produces books on animal advocacy, veganism, religion, social justice, humane education, psychology, family therapy, and recovery. Lantern is dedicated to printing in the United States on recycled paper and saving resources in our day-to-day operations. Our titles are also available as ebooks and audiobooks.

To catch up on Lantern's publishing program, visit us at www.lanternpm.org.

 facebook.com/lanternpm
twitter.com/lanternpm
instagram.com/lanternpm